100
Mental Maths Starters

Contents

Introduction	3
Reference grid	4–5
Starter activities 1–100	6–69
Resource pages	
• Numeral cards 0–10	70
• Numeral cards 11–20	71
• Number track 1–10	72
• Tens number line 0–100	73
• Domino doubles	74
• Number snake 0–20	75
• Make 10 cards	76
• Fact cards: addition	77
• Fact cards: subtraction	78
• 0–99 square	79
Index	80

Margaret Gronow

100 Mental Maths Starters

Year 1

Author Margaret Gronow

Editor Joel Lane

Assistant Editor David Sandford

Cover Design Heather C Sanneh and Clare Brewer

Series Designer Paul Cheshire

Designer Martin Ford

Cover photography Martyn F Chillmaid

Illustrations Louise Gardner

Acknowledgements

The author and publishers wish to thank:

Ann Montague-Smith for permission to base activities on material in her book *100 Maths Lessons and more: Year 1*.

The Controller of HMSO and the DfEE for the use of extracts from *The National Numeracy Strategy: Framework for Teaching Mathematics* © Crown Copyright. Reproduced under the terms of HMSO Guidance Note 8.

Published by Scholastic Ltd,
Villiers House,
Clarendon Avenue,
Leamington Spa,
Warwickshire CV32 5PR

Printed by Bell & Bain Ltd, Glasgow

© **Scholastic Ltd 2002**
Text © Margaret Gronow 2002

1234567890   2345678901

British Library Cataloguing-in-Publication Data
A catalogue record for this book is available from the British Library.

ISBN 0-439-01900-1

About the series

100 Mental Maths Starters is a series of six photocopiable teacher's resource books, one for each of Years 1–6. Each book offers 100 mental maths activities, each lasting between 5 and 10 minutes. These activities are ideal to start your daily dedicated maths lesson if you are following the National Numeracy Strategy. Each year-specific book provides mental activities for maths within the guidelines of the NNS *Framework for Teaching Mathematics*. The activities can also be used effectively to meet the needs of Primary 1–7 classes in Scottish schools, or classes in other schools functioning outside the boundaries of the National Numeracy Strategy.

This series provides suitable questions to deliver the 'Oral and Mental Starters' outlined in the lesson plans in the companion series from Scholastic, *100 Maths Lessons and more*. Reference grids are provided (see pages 4–5) to indicate the lesson and page numbers of the associated lesson plans in the relevant *100 Maths Lessons and more* book. However, the series is also wholly appropriate for independent use alongside any maths scheme of work. The index at the back of each book makes it easy to choose a suitable starter activity for any maths lesson.

Readers of this book who are using *100 Maths Lessons and more: Year 1* will notice that the starters in this book sometimes use different resources or a slightly different method from their counterparts in the companion book. This is intended to provide greater choice and variety, while keeping to a closely similar mathematical content and progression. It is for the teacher to decide when to repeat an activity and when to move on: the exact mix of consolidation and progression needed will vary from one class to another.

Each book provides support for teachers through three terms of mental maths, developing and practising skills that will have been introduced, explained and explored in your main maths lesson time. Few resources are needed, and the questions for each activity are provided in full. The books are complete with answers, ready for you to pick up and use. In addition, all the activities in the book can be photocopied and the answers cut off to leave activity cards that pupils can work from individually. Alternatively, the activity cards can be used by pairs or small groups, with one child asking questions and the other(s) trying to answer.

The activities are suitable for use with single- or mixed-ability groups and single- or mixed-age classes, as much emphasis has been placed on the use of differentiated and open-ended questions. Differentiated questions ensure that all the children can be included in each lesson and have the chance to succeed; suitable questions can be directed at chosen individuals, almost guaranteeing success and thus increased confidence.

Learning mental maths

The mental maths starters in this book provide a structured programme with a balanced progression. They provide regular opportunities for all your children to learn, practise and remember number facts. Completed in order, the activities in this book provide the framework of a scheme of work for mental maths practice in Year 1. Several essential photocopiable

resource pages are also included (see pages 70–9). To cover the whole year, you will need to add some repeats and/or variations of the activities for consolidation. This book does not provide the groundwork concept teaching for each new skill: that is covered in detail, as the focus of appropriately timed main teaching activities in *100 Maths Lessons and more: Year 1* by Ann Montague-Smith.

Each activity in this book has one or more learning objectives based on the 'Teaching Programme: Year 1' in the NNS *Framework*. Key objectives are highlighted in bold. Teacher instructions are provided, stating the particular skills being developed or practised. Discussion of the children's methods is encouraged, since this is essential: it will help the children to develop mathematical language skills: to appreciate that no single method is necessarily 'correct', and that a flexible repertoire of approaches is useful; and to improve their overall confidence as they come to realise that all responses have value. Strategies are encouraged that will enable the children to progress from the known to the unknown number facts, thus developing their ability to select and use methods of mental calculation.

As adults, we probably do maths 'in our heads' more often than we use written methods. Almost without thinking about it, we apply flexible strategies that we developed as children. By following the activities in this series, children will learn to explain their thought processes and techniques – which, in turn, will help them to clarify their thinking and select appropriate methods to use in different contexts.

About this book

This book is aimed at developing the mental and oral skills of pupils in Year 1/Primary 1–2. It builds on the mathematical experiences of children gained during their early years and their work in the Reception year.

In this book, children are helped to develop an understanding of the number system. Much practice is given in counting, sequencing and ordering numbers. Opportunities are provided for children to discuss the patterns that they can see when counting in steps of different sizes (ones, twos, threes, fives and tens). There is a strong emphasis on visual aids such as number tracks, lines and hundred squares; these are crucially important for children in Year 1, at a time when their mathematical experience is limited and their confidence is easily destroyed.

The book provides a programme that helps children to progress from the physical counting of objects to the understanding of the operations of addition and subtraction. Active participation is included where possible to encourage and support the learning of mental strategies. The completion of the work in this book gives a sound basis for work in Year 2 and subsequent years, covered in the later books in the series. By following the activities in these books, children will develop a variety of strategies for the solution of mathematical problems and will learn to be flexible in their approach to numerical work.

This book is, first and foremost, a resource for practising teachers. Comments or suggestions from teachers using the book will thus be very welcome, and may be incorporated into future editions.

Starter activity	Activity	Term	Unit	Lesson	Page
	100 Mental Maths Starters	**100 Maths Lessons**			
1	addition to 5	1	1	1	21
2	subtraction to 5	1	1	2	22
3	ordering numbers to 10	1	2–4	1	26
4	addition to 5	1	2–4	2	27
5	counting to 10	1	2–4	3	28
6	adding by counting on	1	2–4	6	30
7	ordering numbers to 10	1	2–4	8	31
8	addition to 10 by counting on	1	2–4	9	32
9	addition to 10 by counting on	1	2–4	10	32
10	counting reliably a number of objects	1	2–4	11	33
11	addition to 10	1	2–4	12	33
12	addition to 5	1	2–4	13	33
13	counting to 20 and back	1	2–4	14	34
14	counting to 20 and back	1	2–4	15	34
15	ordering numbers to 20	1	5–6	1	40
16	addition doubles of numbers to 5	1	5–6	2	41
17	counting to 20	1	5–6	3	42
18	counting to 20 and back	1	5–6	5	43
19	counting on and back from a small number	1	8	2	52
20	counting on and back from a small number	1	8	3	52
21	pairs totalling 10	1	8	5	53
22	reading numbers to 10	1	9–11	1	57
23	addition to 20 by counting on	1	9–11	2	58
24	addition to 20 by counting on	1	9–11	4	60
25	counting in tens to 100	1	9–11	5	60
26	pairs totalling 10	1	9–11	7	62
27	addition and subtraction to 5	1	9–11	9	63
28	adding by partitioning into '5 and a bit'	1	9–11	11	64
29	pairs totalling 10	1	9–11	14	65
30	addition doubles of numbers to 5	1	12–13	1	73
31	counting reliably a number of objects	1	12–13	2	74
32	counting in tens to 100 and back	1	12–13	4	75
33	pairs totalling 10	1	12–13	8	77
34	addition doubles of numbers to 5	2	1	1	87
35	addition doubles of numbers to 5	2	2–4	1	92
36	counting reliably a number of objects	2	2–4	3	93
37	adding 1, adding 10	2	2–4	4	93
38	ordering tens to 100	2	2–4	5	95
39	pairs totalling 10	2	2–4	8	96
40	adding by partitioning into '5 and a bit'	2	2–4	10	98
41	pairs totalling 10	2	2–4	13	100
42	addition to 5	2	5–6	1	104
43	subtraction to 5	2	5–6	4	106
44	addition doubles of numbers to 10	2	5–6	5	107
45	pairs totalling 10	2	8	1	118
46	counting in twos: odds and evens	2	8	4	120
47	counting to 20 and back	2	9–10	1	124
48	counting reliably a number of objects	2	9–10	2	125
49	addition to 5	2	9–10	4	126
50	subtraction to 5	2	9–10	5	126

Starter activity	Activity	Term	Unit	Lesson	Page
	100 Mental Maths Starters		**100 Maths Lessons**		
51	counting on and back in tens	2	9–10	7	128
52	addition of three numbers	2	9–10	9	130
53	addition of three numbers	2	9–10	10	130
54	addition and subtraction to 10	2	11–12	1	136
55	addition and subtraction to 10	2	11–12	4	138
56	addition and subtraction to 10	2	11–12	7	139
57	addition to 20	3	1	1	149
58	addition to 20	3	1	3	151
59	partitioning a teens number (TU)	3	2–4	1	156
60	partitioning a teens number (TU)	3	2–4	2	156
61	counting on and back in tens	3	2–4	3	157
62	counting on and back in tens	3	2–4	4	158
63	addition to 10	3	2–4	5	159
64	subtraction to 10	3	2–4	6	159
65	addition and subtraction to 10	3	2–4	7	160
66	partitioning a 'teens' number (TU)	3	2–4	10	161
67	partitioning a 'teens' number (TU)	3	2–4	11	162
68	addition to 10	3	2–4	12	162
69	addition and subtraction to 10	3	2–4	13	162
70	ordering numbers to 20: odds and evens	3	2–4	14	163
71	ordering numbers to 20: odds and evens	3	2–4	15	163
72	adding by bridging through 10	3	5–6	1	170
73	pairs totalling 10	3	5–6	2	171
74	adding by bridging through 10	3	5–6	3	172
75	pairs totalling 10	3	5–6	2	171
76	addition doubles of numbers to 5	3	5–6	5	173
77	addition doubles and near doubles of numbers to 5	3	5–6	6	173
78	pairs totalling 10	3	8	1	180
79	counting in ones, twos, tens	3	8	2	181
80	counting in twos: odds and evens	3	8	3	182
81	counting in twos: odd and evens; counting in fives	3	8	4	182
82	counting in threes	3	8	5	182
83	ordering in ones and tens	3	9–11	1	186
84	subtraction to 10	3	2–4	14	163
85	ordering in twos: odd numbers	3	2–4	15	163
86	counting on and back in tens	3	9–11	4	188
87	addition to 10	3	9–11	6	189
88	addition to 10	3	9–11	7	189
89	subtraction to 10	3	9–11	8	191
90	subtraction to 10	3	9–11	9	191
91	counting in fives	3	9–11	10	192
92	counting in fives	3	9–11	11	192
93	counting in threes	3	9–11	12	192
94	addition doubles of numbers to 5	3	9–11	13	193
95	addition doubles of numbers to 5	3	9–11	14	193
96	addition to 10	3	12–13	3	202
97	addition to 10	3	12–13	3	202
98	time: reading o'clock and half past	3	12–13	4	203
99	time: reading o'clock and half past	3	12–13	5	203
100	adding by bridging through 10 and 20	3	12–13	8	205

Answers

1. 3
2. 4
3. 4
4. 4
5. 4
6. 5
7. 5
8. 2
9. 5
10. 5

Starter activity 1

Resources
Two PE hoops; a selection of objects (for example blocks, yoghurt pots, large fir cones); numeral cards 1–5 (enlarged from photocopiable page 70).

Objective
Know by heart addition facts for all pairs of numbers with a total up to at least 5.

Strategies
• Ask one child to put three objects into one hoop. Ask another child to put two objects into the other hoop. Ask: *How many are there altogether?* All count 3 and 2 together, then count 2 and 3 together. Show the card for 5.
• Emphasise that the last number in the count is the number of objects altogether.

Add up

1.	2 and 1		6.	1 and 4
2.	1 and 3		7.	5 and 0
3.	4 and 0		8.	1 and 1
4.	3 and 1		9.	4 and 1
5.	2 and 2		10.	2 and 3

Starter activity 3

Resources
Numeral cards 1–10 (enlarged from photocopiable page 70).

Objective
Count on and back in ones from any small number.

Strategies
• Give the cards 1–5 to a small group of children. Ask them to build a number line. Ask: *Is their number line right?* Count together forwards and backwards. Repeat using different children.
• Give out the cards 1–10. Call out the child holding '1' to start a number line. Ask the other children to build it. When it is correct, count together forwards and backwards. Repeat using different children.

Counting to 10

1. Which number comes after 4?

2. Which number comes after 7?

3. Which number comes before 3?

4. Which number comes before 9?

5. Tell me a number that is bigger than 2.

6. Tell me a number that is bigger than 5.

SCHOLASTIC

Take away

Ask a child to put four objects into the hoop. Ask another child or the children to check that there are four objects. Ask another child to take away one object. Ask: *How many are left?* Count together.

Encourage the children to give a subtraction sentence. All say together: '4 take away 1 is 3.'

1.	3 take away 1	9.	2 take away 1
2.	4 take away 2	10.	5 take away 5
3.	2 take away 1	11.	5 take away 3
4.	5 take away 1	12.	4 take away 1
5.	2 take away 2	13.	2 take away 0
6.	5 take away 2	14.	4 take away 2
7.	3 take away 3	15.	3 take away 2
8.	4 take away 0		

Starter activity 2

Resources
One PE hoop; a selection of objects (eg bricks, yoghurt pots, large fir cones); large numeral cards 0–5 from photocopiable page 70.

Objective
Know by heart addition facts for all pairs of numbers with a total up to at least 5, and the corresponding subtraction facts.

Strategies
● For each question, repeat the instructions used in the first example.

Answers

1. 2
2. 2
3. 1
4. 4
5. 0
6. 3
7. 0
8. 4
9. 1
10. 0
11. 2
12. 3
13. 2
14. 2
15. 1

Answers

11. 3
12. 5
13. 6
14. 7
15. 5

Starter activity 4

Objective

Know by heart addition facts for all pairs of numbers with a total up to at least 5.

Finger add

Explain that when you say a number, you want the children to hold up that number of fingers on one hand. (If necessary, demonstrate touching and raising one finger at a time from a folded hand.)

Show me...

1.   3 fingers
2.   4 fingers
3.   5 fingers
4.   2 fingers
5.   0 fingers

Tell the children that they should now show the first number with one hand and the second number with the other hand. *Show me...*

6.   3 fingers and 1 finger
7.   5 and 2
8.   1 and 4
9.   2 and 0
10.   3 and 5

Explain that you are going to ask them: *How many fingers altogether?* They should use both hands to help them answer.

11.   2 and 1
12.   3 and 2
13.   1 and 5
14.   4 and 3
15.   5 and 0

Number line count

Starter activity 5

Count together from 1 to 10. Count round the class from 1 to 10. Count backwards together from 10 to 1. Count round the class from 10 to 1.

Ask individuals to touch numbers on the number line:

Resources
A class number line showing 1–10.

Objective
Know the number names and recite them in order to at least 10, from and back to zero. **Count on and back in ones from any small number.**

1. the number 4

2. the number that comes after 4

3. the number that comes before 4

4. the number 6

5. the number that comes after 6

6. any number bigger than 6

7. a number that is smaller than 6

8. the number 3

9. the number that comes after 3

10. a number that is smaller than 3

11. the number 9

12. the number that comes before 9

13. the number that is 1 more than 9

14. the number 5

15. the number that is 1 more than 5

SCHOLASTIC

Show me add

1. 4
2. 7
3. 5
4. 5
5. 7
6. 3
7. 6
8. 6
9. 4
10. 3
11. 5
12. 5

Starter activity 6

Resources
Numeral cards 1–5 for each child (enlarged from photocopiable page 70).

Objective
Put the larger number first and count on in ones.

Strategies
• Ask the children to put their cards in order. Count forwards and backwards together. Encourage the children to touch each card as they say the number.
• Say two numbers. The children move the cards with those numbers towards themselves and put a finger on the larger number. When you say *Show me*, they hold it up.

1. 3 and 1
2. 2 and 5
3. 1 and 4
4. 3 and 2
5. 4 and 3
6. 1 and 2

7. 4 and 2
8. 1 and 5
9. 1 and 3
10. 1 and 2
11. 4 and 1
12. 2 and 3

Starter activity 7

Resources
A class number line; large numeral cards 1–10.

Objective
Know the number names and recite them in order to at least 20, from and back to zero. **Count on and back in ones from any small number.**

Strategies
• Start by counting together from 1–10. Repeat. Count together from 10–1. Repeat.
• Deal out the number cards 1–10 to ten children.

Counting with cards

Say: *Come here, number 2.* Ask the children to build a number line, working in both directions from 2. Ask: *Is their number line correct?* Count together forwards and backwards.

Repeat with different children, starting with:

1. 5
2. 8
3. 6

4. 9
5. 4
6. 3

Number track

Ask the children to put a finger on the number 2, using their left hand. Tell them: *Together we are going to count on 3 with a finger from the other hand.* The children say: '1, 2, 3.' Ask: *What number have you landed on?* Say: *2 add 3 equals 5.* The children repeat it with you.

Starter activity 8

Resources
A number track 1–10 (from photocopiable page 72) for each child.

Objective
Put the larger number first and count on in ones.

Answers

1. 5
2. 3
3. 7
4. 4
5. 6
6. 8
7. 8
8. 5
9. 6
10. 6
11. 8
12. 8
13. 9
14. 9

1.   start on 1, count on 4
2.   2   1
3.   5   2
4.   1   3
5.   3   3

6.   2   6
7.   5   3
8.   4   1
9.   1   5
10. 5   1

After the last two questions, ask: *Which was easier?* Ask the children to explain.

11. start on 1, count on 7
12. 7   1

13. 3   6
14. 6   3

Conclude that it is easier to count on from the larger number.

Number track addition

Starter activity 9

Resources
A number track 1–10 (from photocopiable page 72) for each child.

Objective
Put the larger number first and count on in ones.

Strategies
● Remind the children that when adding, it is easier to count on from the larger number. Ask: *Which is the larger number, 4 or 2?* The children put a finger on 4 and count on 2. Say together: '4 add 2 equals 6.'

Each time, ask which is the larger number. Encourage the children to put a finger on this number and count on from it with the other hand. They should raise a hand to give the answer.

1.   3 and 5

2.   6 and 2

3.   1 and 3

4.   2 and 4

5.   3 and 6

6.   4 and 1

7.   5 and 2

8.   5 and 1

9.   2 and 6

10.   2 and 3

11.   5 and 4

12.   4 and 3

Starter activity 10

Resources
Six cubes or other counting items per child.

Objective
Count reliably at least 10 objects.

Strategies
● Ask the children to count six cubes. Count together. Ask them to put the cubes in a straight line. *How many do you have?* Ask them to count from the other end. *How many do you have now?*
● Ask the children to put their cubes in different positions and places. Help them to see that the number stays the same.

Counting cubes

Ask the children to do the following and count the cubes each time.

1.   spread out the cubes on their table

2.   make a tower

3.   make a snake

4.   hide them under their hands

5.   put them in a circle

6.   hide them under their hands, then close their eyes

7.   walk round the room and count them when they get back

8.   jump up and down four times, then count them again

■SCHOLASTIC

Numeral card addition

Starter activity 11

Resources
Numeral cards 2–8
(enlarged from
photocopiable page 70).

Objective
Know by heart addition
facts for all pairs of
numbers with a total up to
at least 5.

Answers
1. 3
2. 5
3. 2
4. 6
5. 4
6. 7
7. 3
8. 6
9. 8
10. 5

The children sit in a circle with the numeral cards spread out on the carpet. Ask them to add 1 in their heads to the number that you say (eg 4). Ask an individual to hold up the relevant numeral card. Together say: '4 add 1 makes 5.' Repeat for:

1. 2
2. 4
3. 1
4. 5
5. 3

Ask the children to add 2 to the number that you say. Encourage them to hold the number in their heads and add 2 mentally.

6. 5
7. 1
8. 4
9. 6
10. 3

Number line addition

1. 3
2. 5
3. 4
4. 4
5. 5
6. 3
7. 2
8. 5
9. 2
10. 5

Starter activity 12

Resources
A class number line.

Objective
Know by heart addition facts for all pairs of numbers with a total up to at least 5.

Strategies
● Ask the children to add together two numbers that you say. They can use the fingers on each hand for the two numbers.
● Ask individuals to say and point to the number on the line. Each time, say the whole statement together (eg '3 add 1 makes 4').

1. 2 add 1
2. 4 add 1
3. 1 add 3
4. 2 add 2
5. 1 add 4

6. 3 add 0
7. 1 add 1
8. 5 add 0
9. 2 add 0
10. 3 add 2

Counting to 20

Starter activity 13

Objective
Count on and back in ones from any small number.

The children sit in a circle and count with you from 1 to 10 and back again, then from 1 to 20 and back again.

Count around the circle from 1 to 10, then back again, until everyone has had two turns. Repeat for 1 to 20 and back again.

Now the children clap to applaud each other!

SCHOLASTIC

Counting to 20

The children sit in a circle and count with you from 1 to 20 and back again.

Tell the children that they have to count to 20 and back from a small number (eg 5). Ask a child to hold the '5' card up for everyone to see.

Repeat, counting from other small numbers:

7   3   8   4   2

Starter activity 14

Resources
Numeral cards 1–10 (enlarged from photocopiable page 70).

Objective
Count on and back in ones from any small number.

Numbers on a line

Explain that one end of the line represents 1 and the other 10.

Shuffle the cards 1–10 and place them face down. Ask a child to pick a card and peg it on the line in a sensible position. Do the other children agree with its position? Discuss this. Repeat for the other cards. When the line is complete, count along it in both directions together.

Repeat the game or progress to cards 11–20, explaining that one end of the line represents 11 and the other 20.

Starter activity 15

Resources
A washing line (or string), pegs, numeral cards 1–20 (enlarged from photocopiable pages 70 and 71).

Objective
Know the number names and recite them in order to at least 20, from and back to zero.

Doubles

Answers

1. 4
2. 8
3. 6
4. 0
5. 10
6. 2

Starter activity 16

Resources
A set of 'Domino doubles' cards (enlarged from photocopiable page 74).

Objective
Know by heart addition doubles of all numbers to at least 5 (eg 4 + 4).

Shuffle the cards and place them face down on the carpet. Hold up one card for the children to see. Ask them what they can see ('both sides are the same', 'there are three dots and three dots', 'there are six dots' and so on). Discuss their answers and use the word 'double'.

Ask a child to choose a card and hold it up for all to see. Say together (for example): '3 add 3 makes 6. Double 3 makes 6.' Repeat with the other dominoes.

Finish by asking questions. The children use their fingers to show the answers.

1. double 2
2. double 4
3. double 3
4. double 0
5. double 5
6. double 1

Number snake

	Start number	Finish number		Start number	Finish number
1.	0	10	6.	17	15
2.	0	20	7.	18	2
3.	0	12	8.	5	12
4.	0	17	9.	2	9
5.	20	10	10.	9	19

Counting on

	Start at	Count on		Start at	Count on
1.	2	3	6.	7	3
2.	1	5	7.	4	6
3.	4	2	8.	3	4
4.	6	3	9.	5	1
5.	8	2	10.	9	2

Counting on (addition)

Answers

1. 9
2. 8
3. 8
4. 4
5. 8

6. 3
7. 6
8. 6
9. 6
10. 5

Starter activity 19

Resources
A number snake 0–20 (from photocopiable page 75) for each child.

Objective
Describe and extend number sequences: **count on and back in ones from any small number.**

Together count from 0 to 20 and back in ones, with the children touching each number on the snake as they say it. Repeat.

After each of the following, ask: *What number have we finished on?*

1. Start on 5, count on 4.
2. Start on 3, count on 5.
3. Start on 6, count on 2.
4. Start on 1, count on 3.
5. Start on 4, count on 4.

Ask the children to turn their snakes over and find the answers to the following questions in their heads. Encourage them to hold the first number in their heads and count on in ones.

6. Start on 2, count on 1.
7. Start on 4, count on 2.
8. Start on 3, count on 3.
9. Start on 1, count on 5.
10. Start on 3, count on 2.

Counting on (subtraction)

Count on from...

1. 3 to 8

2. 5 to 10

3. 2 to 4

4. 1 to 7

5. 4 to 9

6. 6 to 11

7. 9 to 14

8. 7 to 16

9. 4 to 10

10. 8 to 12

Starter activity 20

Resources
A number snake 0–20 (from photocopiable page 75) for each child.

Objective
Count on and back in ones from any small number.

Strategies
● Ask the children to count on each time from the first number to the second in order to find out how many 'steps' they have taken. They must put a finger on the first number they have said and count on as they move the finger to the second number.
● After each question, ask: *How many did you count on?* or *How many steps did you take?*

Answers

1. 5
2. 5
3. 2
4. 6
5. 5
6. 5
7. 5
8. 9
9. 6
10. 4

Finger subtract

Ask the children to show ten fingers, then six fingers. *How many more fingers are needed to make 10?* Count together from 6 to 10.

Ask the children to use their fingers to show you the other number needed to make 10: *How many from... to make 10?*

1. 5

2. 3

3. 8

4. 4

5. 7

6. 1

Starter activity 21

Objective
Count on and back in ones from any small number.

Answers

1. 5
2. 7
3. 2
4. 6
5. 3
6. 9

Starter activity 22

Resources
A board or flip chart; numeral cards 0–10 (enlarged from photocopiable page 70) for each child.

Objective
Read and write numerals from 0 to at least 20.

Strategies
• Ask the children to put their cards in order on the table.
• Ask them to hold up the card with the answer. Choose an individual to write each answer on the board.

Numbers to 20

Show me:

1. 8

2. the number 1 more than 8

3. the number 1 less than 8

4. 3

5. the number 6

6. the number 2 more than 6

7. 7

8. the number 1 less than 7

9. the number 1 more than 4

10. the number 1 less than 9

Answers

1. 12
2. 17
3. 4
4. 18
5. 17
6. 7
7. 4
8. 14
9. 14
10. 18

Starter activity 23

Resources
A number snake 0–20 (from photocopiable page 75) for each child.

Objective
Put the larger number first and count on in ones, including beyond 10 (eg 7 + 5).

Strategies
• Ask the children to add two numbers by putting a finger on the larger number and counting on by the smaller number. They should wait to answer together when you say 'Now'.

Put the larger number first

1. 4 and 8

2. 10 and 7

3. 1 and 3

4. 5 and 13

5. 6 and 11

6. 2 and 5

7. 3 and 1

8. 10 and 4

9. 8 and 6

10. 6 and 12

SCHOLASTIC

Put the larger number first

1. 2 + 3

2. 6 + 2

3. 5 + 5

4. 4 + 6

5. 8 + 1

6. 2 + 9

7. 1 + 7

8. 5 + 4

9. 7 + 5

10. 3 + 3

Starter activity 24

Answers

1. 5
2. 8
3. 10
4. 10
5. 9
6. 11
7. 8
8. 9
9. 12
10. 6

Resources
A board or flip chart, a class number line.

Objective
Put the larger number first and count on in ones, including beyond 10 (eg 7 + 5).

Strategies
• Make sure all the children can see the number line. Write each addition question on the board and decide together which is the larger number. Count on together by the required number of 'steps', using the number line to help.
• A volunteer can write the answer on the board.

Counting tens

1. What comes after 20 on our line?

2. What comes after 60 on our line?

3. What comes after 10 on our line?

4. What comes after 70 on our line?

5. What comes after 90 on our line?

6. What comes before 40 on our line?

7. What comes before 80 on our line?

8. What comes before 30 on our line?

9. What comes before 100 on our line?

10. What comes before 50 on our line?

Starter activity 25

Answers

1. 30
2. 70
3. 20
4. 80
5. 100
6. 30
7. 70
8. 20
9. 90
10. 40

Resources
A tens number line 0–100 (from photocopiable page 73) for each child.

Objective
Count in tens from and back to zero.

Strategies
• Count together in tens from 0 to 100 and back again, with the children pointing to each number as it is said.
• Children raise a hand to answer each question.

Answers

1. 7
2. 2
3. 9
4. 3
5. 5
6. 8
7. 6
8. 10
9. 4
10. 0

Starter activity 26

Resources
A board or flip chart.

Objective
Know by heart all pairs of numbers with a total of 10 (eg 3 + 7).

How many more?

Draw a number line:

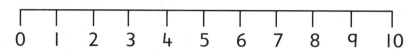

Give the children a number (eg 6) and ask them to find the other number of the pair to make 10. Suggest that they count on from the given number. Ask: *How many steps to 10?*

Point to the numbers as the children count in ones to 10, eg '6 → 1, 2, 3, 4.' Say: *6 add 4 equals 10.* Repeat this number fact together.

1. 3
2. 8
3. 1
4. 7
5. 5

6. 2
7. 4
8. 0
9. 6
10. 10

✦ SCHOLASTIC

Show me add and subtract

1. 2 + 2

2. 3 + 1

3. 1 + 4

4. 5 + 0

5. 2 + 3

6. 1 + 1

7. 2 + 1

8. 3 + 2

9. 5 – 1

10. 3 – 2

11. 4 – 0

12. 2 – 1

13. 5 – 0

14. 1 – 1

15. 3 – 1

16. 2 – 0

Starter activity 27

Objective
Know by heart addition facts for all pairs of numbers with a total up to at least 5, and the corresponding subtraction facts.

Strategies
● Ask the children to show the answers on their fingers. Encourage recall of the number facts.

Answers

1. 4
2. 4
3. 5
4. 5
5. 5
6. 2
7. 3
8. 5
9. 4
10. 1
11. 4
12. 1
13. 5
14. 0
15. 2
16. 2

Pairs to make 10

How many do you need to add to... to make 10?

1. 6

2. 3

3. 8

4. 5

5. 2

6. 1

7. 7

8. 10

9. 0

10. 4

Starter activity 29

Resources
A number track 1–10 (from photocopiable page 72) for each child.

Objectives
Know by heart: **all pairs of numbers with a total of 10** (eg 3 + 7); addition facts for all pairs of numbers with a total up to at least 5, and the corresponding subtraction facts.

Strategies
● Ask: *How many do we need to add to 7 to make 10?* Count on together from 7. Children raise hands to answer.
● Encourage the children to touch the number and count the 'steps' to 10.

Answers

1. 4
2. 7
3. 2
4. 5
5. 8
6. 9
7. 3
8. 0
9. 10
10. 6

Partitioning

Answers

1. 5 and 3
2. 5 and 1
3. 5 and 4

4. 13
5. 11
6. 14

Starter activity 28

Resources
Nine interlocking cubes for each child; a board or flip chart.

Objective
Begin to partition into '5 and a bit' when adding 6, 7, 8 or 9, then recombine (eg 6 + 8 = 5 + 1 + 5 + 3 = 10 + 4 = 14).

Ask the children to make a tower of seven cubes. Explain that they are going to make '5 and a bit' from this tower. They should count 5 into one tower, then count how many are left. Ask: *How many more cubes are there than 5?*

Say: *We made 5, and 2 more, from 7.* Repeat this together.

Practise making '5 and a bit' from:

1. 8 cubes

2. 6 cubes

3. 9 cubes

Write 5 + 7 on the board. Remind the children that they can make '5 and a bit' to help them add numbers together. Ask for suggestions. Encourage the response:
5 + 5 + 2 = 10 + 2 = 12

4. 5 + 8

5. 5 + 6

6. 5 + 9

Doubles and near doubles

Starter activity 30

Remind the children that 'double 3' is the same as 3 + 3.

Ask these questions. Children raise a hand to answer.

1. double 2

2. double 4

3. double 5

4. double 1

5. double 3

Resources
A board or flip chart.

Objectives
Know by heart addition doubles of all numbers to at least 5 (eg 4 + 4). Identify near doubles, using doubles already known (eg 6 + 5).

1. 4
2. 8
3. 10
4. 2
5. 6
6. 5
7. 9
8. 3
9. 11
10. 7

Write:

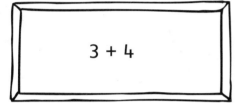

3 + 4

Ask for the answer. Ask several children how they worked it out. Encourage using a 'near double', in this case 3 + 3 + 1.

6. 2 + 3

7. 5 + 4

8. 1 + 2

9. 6 + 5

10. 4 + 3

Counting

Answers

1. 9
2. 8
3. 4
4. 6
5. 9
6. 2
7. 0
8. 8
9. 5
10. 9

Starter activity 31

Resources
Nine cubes for each child.

Objective
Count reliably at least 10 objects.

Strategies
● Ask the questions in sequence. Children raise a hand to answer.

1. Count out 9 cubes.

2. Take one away. How many now?

3. Take away another 4. How many do you have now?

4. Add 2 cubes. Count the cubes.

5. Add another 3 cubes. How many now?

6. Take 7 away. How many are left?

7. Take 2 away. How many do you have?

8. Add 8 cubes. Count the cubes.

9. Take away 3. How many do you have?

10. Add 4. How many do you have now?

Counting tens

Answers

1. 50
2. 70
3. 90
4. 70
5. 90
6. 50
7. 30
8. 20
9. 10
10. 40

Starter activity 32

Resources
A tens number line 0–100 (from photocopiable page 73) for each child.

Objective
Describe and extend number sequences: **count in tens from and back to zero.**

Strategies
● The children point to each number as they count in tens from 0 to 100 and back.
● Ask them to start at 30 and count on 4 tens. Ask: *Where have you counted to?*
● For questions 6–10, explain to the children that they are going to take tens away. Stress that the answer will be a smaller number.

1. Start at 20, count on 3 tens.

2. Start at 50, count on 2 tens.

3. Start at 80, count on 1 ten.

4. Start at 30, count on 4 tens.

5. Start at 60, count on 3 tens.

6. Start at 70, take 2 tens away.

7. Start at 40, take 1 ten away.

8. Start at 90, take 7 tens away.

9. Start at 60, take 5 tens away.

10. Start at 80, take 4 tens away.

SCHOLASTIC

Cards to make 10

Explain that you want to know how many more to 10. The children with the correct card hold it up and say the number.

Hold up the teacher's '6' card and say: *6 add how many equals 10?* The children with a '4' card should hold it up. Say: *6 add 4 equals 10.* Repeat this number fact together.

1. 8 + how many = 10?   6. 2 + how many = 10?

2. 5 + how many = 10?   7. 7 + how many = 10?

3. 9 + how many = 10?   8. 1 + how many = 10?

4. 3 + how many = 10?   9. 6 + how many = 10?

5. 4 + how many = 10?   10. 0 + how many = 10?

Starter activity 33

Resources
Four or more sets of numeral cards 1–9 (enlarged from photocopiable page 70), including a set for the teacher.

Objective
Know by heart all pairs of numbers with a total of 10 (eg 3 + 7).

Strategies
● Give each child one numeral card. More able children could have an extra card.

Answers
1. 2
2. 5
3. 1
4. 7
5. 6
6. 8
7. 3
8. 9
9. 4
10. 10

Starter activity 34

Resources
A set of 'Domino doubles' cards (enlarged from photocopiable page 74).

Objective
Know by heart addition doubles of all numbers to at least 5 (eg 4 + 4).

Double it

Explain that you will be asking the children to double numbers. Remind them that 2 + 2 = double 2 = 4.

Ask a child to pick a domino card from the set (eg double 3) and hold it up for everyone to see. Ask for a volunteer to say what this double makes: 'Double 3 is 6.' Repeat this together.

Repeat with a different child, until all the cards have been used at least twice.

Starter activity 35

Resources
A set of numeral cards 0–10 (enlarged from photocopiable page 70); a washing line (or string), pegs; a set of 'Domino double' cards (enlarged from photocopiable page 74).

Objective
Know by heart addition doubles of all numbers to at least 5 (eg 4 + 4).

Double it

Give out the numeral cards. Place the domino cards face down. Ask a child who is not holding a numeral card to choose a domino card (eg double 3) and hold it up for all to see.

The child with the matching numeral card ('6') stands next to the first child. They say together: 'Double 3 is 6.' The class repeat this.

Continue until all the domino cards have been used at least twice.

Ask the children whose numeral cards have not been used to peg their cards on the line in order. Tell the children that these numbers are called 'odd numbers'. Say them together: 1, 3, 5, 7, 9.

Counting

Ask the children to count out 10 cubes. Count them together.

Say: *Put them in a straight line. How many do you have now? Count them from the other end. How many are there?*

Ask the children to put their 10 cubes in a different order to reinforce the fact that the number will stay the same unless some cubes are added or taken away.

Ask them to do the following and count the cubes each time.

1. spread out the cubes on their table

2. make a snake

3. put the cubes in a circle

4. walk round the table and count the cubes when they get back

5. make a tower

6. jump up and down five times

7. clap their hands four times

8. push the cubes together into a group.

Resources
10 cubes for each child.

Objective
Count reliably at least 10 objects.

SCHOLASTIC

Answers

Starter activity 37

Resources
A 0–99 square (from photocopiable page 79) for each child.

Objective
Within the range 0–30, say the number that is 1 or 10 more than any given number.

Strategies
• After the first three, discuss the pattern.

1 or 10 more

Together, touch and count the numbers from zero to 30 and back again.

Add 1 to each number.

1.	3	5.	9
2.	6	6.	19
3.	16	7.	29
4.	26	8.	12

Add 10 to each number.

9.	5	13.	4
10.	18	14.	9
11.	7	15.	0
12.	13	16.	20

Count in tens

1. What number is 10 more than 20?

2. What number is 10 more than 50?

3. What tens number is between 30 and 50?

4. What tens number is between 60 and 80?

5. What number is 10 less than 90?

6. What number is 10 less than 70?

7. Which number is bigger, 80 or 40?

8. Which is smaller, 90 or 30?

9. Which is bigger, 50 or 10?

10. Which is smaller, 70 or 40?

Starter activity 38

Resources
A tens number line 0–100 (from photocopiable page 73) for each child.

Objective
Describe and extend number sequences: **count in tens from and back to zero.**

Strategies
● Ask the children to point to each number as they count together in tens from zero to 100, then back to zero. Repeat.
● Children raise a hand to answer the questions.

Answers
1. 30
2. 60
3. 40
4. 70
5. 80
6. 60
7. 80
8. 30
9. 50
10. 40

Fast 10

1. 8

2. 4

3. 9

4. 2

5. 0

6. 5

7. 3

8. 6

9. 1

10. 7

Starter activity 39

Resources
A set of numeral cards 0–10 (enlarged from photocopiable page 70) for each child and the teacher.

Objective
Know by heart all pairs of numbers with a total of 10 (eg 3 + 7).

Strategies
● Explain that you will hold up a numeral card and say the number (eg 7). The children will hold up the card for the number that adds to your number to make 10. Encourage them to count on to 10 in their heads.
● Ask an individual to say the answer. Say together: '7 add 3 makes 10.'

Answers
1. 2
2. 6
3. 1
4. 8
5. 10
6. 5
7. 7
8. 4
9. 9
10. 3

Partitioning

Answers

1. 12

2. 14

3. 14

4. 15

Starter activity 40

Resources
A board or flip chart.

Objective
Begin to partition into '5 and a bit' when adding 6, 7, 8 or 9, then recombine (eg 6 + 8 = 5 + 1 + 5 + 3 = 10 + 4 = 14).

Remind the children of the strategy of partitioning larger numbers into '5 and a bit', then recombining to make 10, then counting on. Write: 5 + 8 =

Ask: *How can we do this?* Encourage the partitioning and recombining strategy. Write:

$$5 + 8 = 5 + 5 + 3 = 10 + 3 = 13$$

Emphasise that 5 + 5 = 10, and that adding on to 10 is easy!

Write each example on the board and ask an individual to explain how he or she is going to work it out.

1. 5 + 7

2. 5 + 9

Write 6 + 7 = and ask for suggestions about how to find the answer. Write 5 + 1 + 5 + 2. Encourage adding the fives first to make 10, then adding the ones:

$$6 + 7 = 5 + 1 + 5 + 2 = 10 + 3 = 13$$

3. 8 + 6

4. 7 + 8

Clock 10

Fact cards

Tell the children that you will hold up a card for them to read aloud with you. You will then ask for someone to give the answer. If the others agree, they should give it the 'thumbs up'; if not, they should give it the 'thumbs down'.

Keep the pace as fast as possible. Make a note of how long it takes to work through all the cards. Repeat at a later date (see Starter activity 49).

Fact cards

Tell the children that you will hold up a card for them to read aloud with you. You will then ask for someone to give the answer. If the others agree, they should give it the 'thumbs up'; if not, they should give it the 'thumbs down'.

Keep the pace as fast as possible. Make a note of how long it takes to work through all the cards. Repeat at a later date (see Starter activity 50).

Fast 10

Explain that you will hold up a card with a number sentence on it that totals 10 but has one number missing (eg 7 + $\square$ = 10). Ask the children to read the card with you, saying 'how many?' for the missing number.

Ask the children to hold up a numeral card to show the missing number. Say the fact together: '7 add 3 equals 10.'

Work through all the 'Make 10' cards.

Doubles

Spread the numeral cards face up on the carpet. Ask an individual to choose the card that is the answer to each question and stand with it in front of the class.

Starter activity 44

Resources
A set of numeral cards 1–20 (enlarged from photocopiable pages 70 and 71).

Objective
Know by heart addition doubles of all numbers to at least 5 (eg 4 + 4).

1. What is double 2?

2. 5 add 5

3. double 3

4. 8 add 8

5. double 10

6. double 6

7. What is double 1?

8. double 4

9. 7 add 7

10. double 9

When all the questions have been answered, ask the children who are holding up the cards to stand in order. What do the children notice about these numbers? They may say that 'some are missing' or 'they are every other number'. Explain that these numbers are called 'even numbers'. Say the even numbers from 2 to 20 together.

Answers

1. even
2. odd
3. odd
4. even
5. odd
6. even
7. even
8. odd

Starter activity 46

Resources
A set of numeral cards 0–10 (enlarged from photocopiable page 70).

Objective
Count on in twos from zero, then one, and begin to recognise odd or even numbers to about 20 as 'every other number'.

Odd or even

Give out the numeral cards and build a number line, with each child holding up a card. Count aloud together. Ask the child with the '1' card to kneel, and then every alternate child, until all the odd numbers are kneeling and the evens standing.

Count together in twos from zero. Explain that these are called **even numbers**.

Count together in twos from 1. Explain that these are called **odd numbers**.

Count together from 0 to 10, shouting the odd numbers and whispering the even numbers.

Ask whether the following numbers are odd or even, and how the children know (for example, 'I said "4" when I counted in 2s from 0').

1.	8	5.	1
2.	5	6.	4
3.	3	7.	2
4.	6	8.	7

Counting patterns

The children sit in a circle. Count together from zero to 20, then back to zero. Count around the class, forwards and backwards.

Count together from a small number up to 20 and back to zero. Repeat, this time counting around the class, until everyone has had at least two turns.

Start numbers:

1. 5   3. 7   5. 8

2. 3   4. 4

Starter activity 47

Objective
Describe and extend number sequences: **count on and back in ones from any small number.**

Counting out

1. Make a line with your cubes. Count them.

2. Make two rows of 6. Count the cubes.

3. Hold one cube in a hand. How many are on the table now?

4. Make a circle with all your cubes. Count them.

5. Put one cube in each hand. How many are on the table now?

6. Move 4 cubes away from the others. How many are left?

7. Put 4 cubes in a row, then two more rows of 4. How many cubes are there altogether?

8. Take 5 cubes away. How many are left?

9. Put 3 cubes in a row, then 3 more rows of 3. How many cubes are there altogether?

10. Take 3 cubes away. How many are left?

Starter activity 48

Resources
Twelve cubes for each child.

Objective
Count reliably at least 20 objects.

Strategies
● Children raise a hand to answer each question.

Starter activity 49

Resources
A set of 'Fact cards' for addition (enlarged from photocopiable page 77).

Objective
Know by heart addition facts for all pairs of numbers with a total up to at least 5.

Fact cards

Tell the children that you will hold up a card for them to read aloud with you. You will then ask for someone to give the answer. If the others agree, they should give it the 'thumbs up'; if not, they should give it the 'thumbs down'.

If appropriate, remind the children how long they took before (see Starter activity 42). Can they do it faster this time? Keep the pace as fast as possible. Make a note of how long it takes to work through all the cards.

Starter activity 50

Resources
A set of 'Fact cards' for subtraction (enlarged from photocopiable page 78).

Objective
Know by heart addition facts for all pairs of numbers with a total up to at least 5, and the corresponding subtraction facts.

Fact cards

Tell the children that you will hold up a card for them to read aloud with you. You will then ask for someone to give the answer. If the others agree, they should give it the 'thumbs up'; if not, they should give it the 'thumbs down'.

If appropriate, remind the children how long they took before (see Starter activity 43). Can they do it faster this time?

Keep the pace as fast as possible. Make a note of how long it takes to work through all the cards.

10 more or less

Starter activity 51

Resources
A 0–99 square (from photocopiable page 79) for each child.

Objective
Within the range 0–30, say the number that is 1 or 10 more or less than any given number.

Ask the children to count together in ones from a given start number. They should keep a finger on the start number.

Count on 10 from...

1. 3

2. 5

3. 16

4. 14

5. 10

Count back 10 from...

6. 17

7. 12

8. 18

9. 11

10. 19

Answers

1. 7
2. 8
3. 9
4. 7
5. 8
6. 9
7. 9
8. 9
9. 10
10. 9

Starter activity 52

Resources
Three PE hoops, a selection of objects (eg bricks, yoghurt pots, large fir cones), a board or flip chart.

Objective
Begin to recognise that more than two numbers can be added together.

Strategies
● Place the three hoops on the carpet and ask a child to place two objects in one hoop. Ask others to put three objects in the second hoop and one in the third. Ask: *How many objects are there altogether?*
● Write: 2 + 3 + 1 = 6. Count up together.
● Repeat the activity.

Adding three numbers

1. 1 + 4 + 2
2. 3 + 2 + 3
3. 1 + 5 + 3
4. 2 + 2 + 3
5. 3 + 1 + 4

6. 2 + 5 + 2
7. 1 + 2 + 6
8. 3 + 3 + 3
9. 3 + 5 + 2
10. 4 + 3 + 2

Answers

1. 9
2. 11
3. 10
4. 10
5. 13
6. 13
7. 14
8. 10
9. 16
10. 18

Starter activity 53

Resources
A set of numeral cards 0–10 (enlarged from photocopiable page 70).

Objective
Begin to recognise that more than two numbers can be added together.

Strategies
● Ask three children to each hold up a numeral card. Ask the class to add the three numbers together. Encourage counting on from the first number.
● Start by counting together. As their confidence increases, the children can work individually and raise a hand to answer.

Adding three numbers

1. 4 + 2 + 3
2. 3 + 6 + 2
3. 5 + 1 + 4
4. 8 + 2 + 0
5. 1 + 7 + 5

6. 3 + 2 + 8
7. 9 + 0 + 5
8. 1 + 6 + 3
9. 7 + 9 + 0
10. 4 + 6 + 8

Trios

Starter activity 54

Resources
Numeral cards 0–10 and signs +, – and = (enlarged from photocopiable pages 70 and 71), a board or flip chart.

Objective
Begin to know addition facts for all pairs of numbers with a total up to at least 10, and the corresponding subtraction facts.

Choose five children to hold up the cards:

| 2 | 3 | 5 | + | = |

Ask them to stand in order to make an addition statement (eg 2 + 3 = 5). Ask: *Can we make a different addition statement?* (3 + 2 = 5)

Replace the + sign with a – and ask the children to make a subtraction statement. Ask: *Can we make a different 'take away'?* Discuss why 2 – 5 = 3 is not correct.

Write the four statements on the board and say them together.

Repeat the activity with:

1.   6   2   4

2.   3   5   8

3.   7   4   3

4.   4   9   5

5.   2   8   10

**Starter
activity 55**

Resources
Numeral cards 0–10 and
signs +, – and =
(enlarged from
photocopiable pages 70
and 71), a board or flip
chart.

Objective
Begin to know addition
facts for all pairs of
numbers with a total up to
at least 10, and the
corresponding subtraction
facts.

Trios

Choose five children to hold up the cards:

| 7 | 5 | 2 | + | = |

Ask them to stand in order to make an addition
statement (eg 2 + 5 = 7). Ask: *Can we make a
different addition statement?* (5 + 2 = 7)

Replace the + sign with a – and ask the children to
make a subtraction statement. Ask: *Can we make a
different subtraction statement?*

Write the four statements on the board. Discuss them,
then say them together.

Repeat the activity with:

1.   6   8   2

2.   10   7   3

3.   6   9   3

4.   7   5   2

5.   4   6   10

Addition and subtraction facts

1. 6 + 1

2. 10 + 2

3. 8 – 5

4. 5 – 1

5. 4 + 6

6. 3 + 5

7. 9 + 1

8. 8 – 4

9. 10 – 2

10. 7 – 5

11. 2 + 6

12. 2 + 7

13. 10 – 5

14. 9 – 2

15. 3 + 6

Starter activity 56

Resources
A 0–20 number snake (from photocopiable page 75) for each child.

Objective
Use known number facts and place value to add or subtract a pair of numbers mentally within the range 0 to at least 10, then 0 to at least 20. (See Starter activity 57.)

Strategies
● The children raise a hand to answer.
● Encourage them to recall facts or workings without using the snake.

Answers

1. 7
2. 12
3. 3
4. 4
5. 10
6. 8
7. 10
8. 4
9. 8
10. 2
11. 8
12. 9
13. 5
14. 7
15. 9

Addition beyond 10

1. 10 + 4

2. 10 + 6

3. 10 + 1

4. 10 + 9

5. 10 + 3

6. 10 + 8

7. 10 + 5

8. 10 + 2

9. 10 + 7

10. 12 + 4

11. 12 + 1

12. 12 + 3

Starter activity 57

Resources
A 0–20 number snake (from photocopiable page 75) for each child.

Objective
Use known number facts and place value to add or subtract a pair of numbers mentally within the range 0 to at least 10, then 0 to at least 20.

Strategies
● Ask the children to find the answers using the number snake.
● Count on together from the larger number.

Answers

1. 14
2. 16
3. 11
4. 19
5. 13
6. 18
7. 15
8. 12
9. 17
10. 16
11. 13
12. 15

Photocopiable

Addition beyond 10

Starter activity 58

Resources
A 0–20 number snake (from photocopiable page 75) for each child.

Objective
Use known number facts and place value to add or subtract a pair of numbers mentally within the range 0 to at least 10, then 0 to at least 20.

Strategies
• Ask the children to find the answers using the number snake.
• Count on together from the larger number.

1. $11 + 2$
2. $11 + 5$
3. $11 + 0$
4. $11 + 7$
5. $11 + 1$
6. $11 + 8$

7. $11 + 3$
8. $11 + 9$
9. $11 + 4$
10. $11 + 6$
11. $12 + 4$
12. $12 + 2$

Teens

Starter activity 59

Resources
A board or flip chart.

Objectives
Begin to know what each digit in a two-digit number represents. Partition a 'teens' number into 10 and ones.

Strategies
• Ask the children to find the answer in their heads and raise a hand to answer.
• For each response, ask: *How many tens? How many units?*
• Write the answer on the board, eg '17 = 1 ten and 7 units'. Say it together.

1. $10 + 4$
2. $10 + 8$
3. $10 + 6$
4. $10 + 9$
5. $10 + 3$

6. $10 + 1$
7. $10 + 5$
8. $10 + 0$
9. $10 + 2$
10. $10 + 7$

SCHOLASTIC

Teens

1. 11 + 6
2. 11 + 4
3. 11 + 2
4. 11 + 8
5. 11 + 1

6. 11 + 5
7. 11 + 0
8. 11 + 7
9. 11 + 3
10. 11 + 9

Starter activity 60

Resources
A board or flip chart.

Objectives
Begin to know what each digit in a two-digit number represents. Partition a 'teens' number and begin to partition larger two-digit numbers into a multiple of 10 and ones (TU).

Strategies
● Ask the children to find the answer in their heads and raise a hand to answer.
● For each response, ask: *How many tens? How many units?*
● Write the answer on the board, eg '17 = 1 ten and 7 units'. Say it together.

Answers
1. 17
2. 15
3. 13
4. 19
5. 12
6. 16
7. 11
8. 18
9. 14
10. 20

---

10 more, 10 less

What is 10 more than...?

1. 5
2. 13
3. 20

4. 19
5. 6
6. 17

What is 10 less than...?

7. 28
8. 12
9. 17

10. 20
11. 14
12. 26

Starter activity 61

Resources
A 0–99 square (photocopiable page 79) for each child.

Objective
Within the range 0 to 30, say the number that is 1 or 10 more or less than any given number.

Strategies
● The children should touch each number as they count on (or back). Encourage making a 'step of 10' where appropriate.
● After the first three questions, discuss any patterns the children have noticed.

Answers
1. 15
2. 23
3. 30
4. 29
5. 16
6. 27
7. 18
8. 2
9. 7
10. 10
11. 4
12. 16

10 more, 10 less

Answers

1. 18
2. 20
3. 11
4. 25
5. 28
6. 26
7. 16
8. 5
9. 19
10. 1
11. 3
12. 6

Starter activity 62

Resources
A 0–99 square (photocopiable page 79) for each child.

Objective
Within the range 0 to 30, say the number that is 1 or 10 more or less than any given number.

Strategies
● Explain that you will say a number and will ask the children for a number that is 10 more than it.
● The children should touch each number as they count on (or back). Encourage making a 'step of 10' where appropriate.

What is 10 more than...?

1. 8
2. 10
3. 1

4. 15
5. 18
6. 16

What is 10 less than...?

7. 26
8. 15
9. 29

10. 11
11. 13
12. 16

Addition to 10

Answers

1. 7
2. 9
3. 9
4. 7
5. 9
6. 10
7. 8
8. 10
9. 10
10. 9
11. 9
12. 10

Starter activity 63

Resources
A number track 1–10 (photocopiable page 72) for each child.

Objective
Begin to know addition facts for all pairs of numbers with a total up to at least 10.

Strategies
● Explain that you will ask the children to find the answer to a 'sum' or addition question.
● Encourage them to count on, using the number track. More able children can be encouraged to count on from the larger number.

1. 6 + 1
2. 2 + 7
3. 1 + 8
4. 7 + 0
5. 6 + 3
6. 1 + 9

7. 3 + 5
8. 4 + 6
9. 8 + 2
10. 0 + 9
11. 4 + 5
12. 7 + 3

■SCHOLASTIC

Subtraction to 10

1. 6 – 2

2. 4 – 1

3. 10 – 7

4. 5 – 3

5. 9 – 6

6. 7 – 3

7. 5 – 1

8. 8 – 7

9. 6 – 6

10. 3 – 2

11. 10 – 5

12. 9 – 3

Starter activity 64

Resources
A number track 1–10 (photocopiable page 72) for each child.

Objective
Begin to know addition facts for all pairs of numbers with a total up to at least 10, and the corresponding subtraction facts.

Strategies
• Explain that you will ask the children to find the answer to a subtraction ('take away') question.
• Remind the children that they should 'hop' back from the start number, saying aloud the number of 'hops' that they make.

Answers
1. 4
2. 3
3. 3
4. 2
5. 3
6. 4
7. 4
8. 1
9. 0
10. 1
11. 5
12. 6

Addition and subtraction to 10

1. 5 + 2

2. 8 + 1

3. 7 – 6

4. 4 – 4

5. 2 + 6

6. 8 – 2

7. 6 + 2

8. 2 + 4

9. 8 – 5

10. 6 – 3

11. 9 + 1

12. 7 – 4

Starter activity 65

Resources
A number track 1–10 (photocopiable page 72) for each child.

Objective
Begin to know addition facts for all pairs of numbers with a total up to at least 10, and the corresponding subtraction facts.

Strategies
• Explain that some questions will be about adding and some will be about subtracting ('taking away').
• Ask the children to touch the start number and take 'hops' forward to add, or back to take away.

Answers
1. 7
2. 9
3. 1
4. 0
5. 8
6. 6
7. 8
8. 6
9. 3
10. 3
11. 10
12. 3

Teens

1. 14
2. 16
3. 13
4. 17
5. 15
6. 14
7. 18
8. 15
9. 17
10. 16

Starter activity 66

Resources
A board or flip chart.

Objectives
Begin to know what each digit in a two-digit number represents. Partition a 'teens' number into 10 and ones.

Strategies
● Ask the children to raise a hand to answer. Encourage them to hold the first number in their heads, then count on.
● For each response, ask: *How many tens? How many units?*
● Write the answer on the board, eg '17 = 1 ten and 7 units'.

1. $12 + 2$
2. $12 + 4$
3. $12 + 1$
4. $12 + 5$
5. $12 + 3$

6. $13 + 1$
7. $13 + 5$
8. $13 + 2$
9. $13 + 4$
10. $13 + 3$

Teens

1. 13
2. 13
3. 12
4. 17
5. 18
6. 12
7. 14
8. 18
9. 14
10. 17

Starter activity 67

Resources
A set of numeral cards 10–20 (enlarged from photocopiable pages 70 and 71) for each child, a board or flip chart.

Objectives
Begin to know what each digit in a two-digit number represents. Partition a 'teens' number into 10 and ones.

Strategies
● Ask the children to hold up a numeral card to show the answer. For each response, ask how many tens and how many units.
● Write the answer, eg '17 = 1 ten and 7 units'. All say this together.

Show me...

1. 1 more than 12
2. 1 less than 14
3. the number that is 2 more than 10
4. the number that comes after 16
5. the number that comes before 19

6. 1 more than 11
7. 1 less than 15
8. the number that is between 17 and 19
9. 1 more than 13
10. 1 less than 18

Addition to 10

1. 3 + 1

2. 4 + 1

3. 2 + 3

4. 1 + 2

5. 6 + 1

6. 3 + 3

7. 1 + 4

8. 2 + 2

9. 4 + 0

10. 3 + 4

11. 5 + 2

12. 3 + 0

13. 1 + 3

14. 5 + 3

15. 2 + 4

Starter activity 68

Objective
Begin to know addition facts for all pairs of numbers with a total up to at least 10.

Strategies
● Ask the children to find the answer in their heads (without using a number track, number line etc). Encourage them to hold the first number in their heads and count on to find the answer.

Answers

1. 4
2. 5
3. 5
4. 3
5. 7
6. 6
7. 5
8. 4
9. 4
10. 7
11. 7
12. 3
13. 4
14. 8
15. 6

Addition and subtraction to 10

1. 5 + 1

2. 2 + 6

3. 1 + 1

4. 3 + 2

5. 2 + 1

6. 4 + 2

7. 2 + 5

8. 5 + 4

9. 5 – 2

10. 3 – 0

11. 6 – 1

12. 9 – 5

13. 7 – 2

14. 4 – 4

15. 8 – 6

Starter activity 69

Objective
Begin to know addition facts for all pairs of numbers with a total up to at least 10, and the corresponding subtraction facts.

Strategies
● Ask the children to find the answer in their heads. Encourage them to recall the number facts, or to count on from the first number (or the larger number, if they can).

● Tell the children that they now have to find the answers to 'take aways'.

Answers

1. 6
2. 8
3. 2
4. 5
5. 3
6. 6
7. 7
8. 9
9. 3
10. 3
11. 5
12. 4
13. 5
14. 0
15. 2

Starter activity 70

Resources
A set of numeral cards 0–20 (enlarged from photocopiable pages 70 and 71).

Objective
Count on and back in ones from any small number, and in tens from and back to zero; count in twos from zero, then one, and begin to recognise odd or even numbers to about 20 as 'every other number'.

Counting

Deal out the numeral cards 0–10. Say: *Come here, number 4.* Ask the children with cards to build a 0–10 number line in both directions from the 4.

Give out the numeral cards 11–20. Say: *Come here, number 15.* Ask the children to build an 11–20 number line in both directions.

All count together from 0 to 20. Explain that 'every other child' is to sit down, beginning with '1'. All say: 'Sit, stand, sit, stand...'.

Count on in twos from zero. Tell the children that these are the **even** numbers.

Repeat the activity with the odd numbers standing.

Counting

Deal out the numeral cards 0–10. Explain that only odd numbers are allowed on the washing line. Ask the children to hold up their number if they think it is an odd one. Point to a child, who says his or her number. The class decide whether the number should be pegged on the line. Repeat for each number that is held up.

When the numbers are in order, count in twos together from 1 to 9.

Repeat the activity with the numbers 11–20. Now ask the children holding the even numbers to make a standing number line. Count in twos together from 0 to 20 and back again.

Starter activity 71

Resources
A set of numeral cards 0–20 (enlarged from photocopiable pages 70 and 71), a washing line or string, pegs.

Objective
Count on and back in ones from any small number, and in tens from and back to zero; count in twos from zero, then one, and begin to recognise odd or even numbers to about 20 as 'every other number'.

Bridging 10

Write:

Ask for the answer and how the children worked it out. Collect several methods. Encourage making 10:

$6 + 7 = 6 + 4 + 3 = 10 + 3 = 13$

Write each example and ask an individual how he or she is going to find the answer.

1. $4 + 8$

2. $7 + 9$

3. $7 + 6$

4. $5 + 7$

5. $9 + 5$

Starter activity 72

Resources
A board or flip chart.

Objective
Begin to bridge through 10 when adding a single-digit number.

Strategies
● Remind the children of the strategy for adding two numbers when the answer will cross through 10. Stress that they are looking to make 10 because adding on to 10 is easy!

Answers

1. 12

2. 16

3. 13

4. 12

5. 14

Make 10

Explain that you will hold up a card with a number sentence on it that totals 10, but has one number missing (eg 7 + ☐ = 10). Ask the children to read the card with you, saying 'How many?' for the missing number. They should hold up a numeral card to show their answer.

Say the number sentence together, for example '7 add 3 equals 10'.

Bridging 10

Answers

1. 12
2. 13
3. 14
4. 13
5. 12
6. 13
7. 16
8. 11

Write each example on the chart. Ask an individual to give the answer and explain how he or she worked it out.

1. 7 + 5
2. 5 + 8
3. 8 + 6
4. 6 + 7

5. 9 + 3
6. 8 + 5
7. 7 + 9
8. 6 + 5

Make 10

Ask the children to find as many ways as possible of making 10 by breaking their 'tower' of 10 into two parts (eg 4 and 6).

Write on the board:

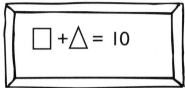

Ask volunteers to give pairs of numbers that could replace the symbols, writing on the board.

When all the possibilities have been collected (including inverse examples such as 4 + 6 and 6 + 4), copy them out in order. Discuss the patterns.

Starter activity 75

Resources
A tower of 10 interlocking cubes for each child; a board or flip chart.

Objectives
Know by heart all pairs of numbers with a total of 10 (eg 3 + 7). Begin to recognise the use of symbols such as □ or △ to stand for an unknown number.

Doubles Snap

Explain that you are going to show the children a 'Domino double' (eg double 2). The children who have the number that equals the total for this domino (ie 4) can hold it up and say 'Snap'.

Continue until all the 'Domino doubles' cards have been used at least three times.

Starter activity 76

Resources
One set of 'Domino doubles' cards (enlarged from photocopiable page 74); enough sets of numeral cards 0–10 (enlarged from photocopiable page 70) for each child to be given an 'even number' card.

Objective
Know by heart addition doubles of all numbers to at least 5 (eg 4 + 4).

Doubles and near doubles

Answers

1. 4
2. 8
3. 0
4. 2
5. 10
6. 6
7. 5
8. 9
9. 7
10. 11

Starter activity 77

Objectives
Know by heart addition doubles of all numbers to at least 5 (eg 4 + 4). Identify near doubles, using doubles already known (eg 6 + 5).

Ask the children to think of the 'Domino doubles' cards (see Starter activity 76) to help them find the answers in their heads.

1. double 2

4. double 1

2. double 4

5. double 5

3. double 0

6. double 3

Suggest that to add 3 and 4, the children could find double 3 and then add 1 (3 + 3 + 1).

7. 2 + 3

9. 3 + 4

8. 4 + 5

10. 5 + 6

Make 10

Explain that you will hold up a numeral card (eg 6) and say, *I have 6*. The children will then find the number that, when added to 6, will make 10. They hold up this card. Point to an individual, who will say: 'I have 4'.

Say together: '6 plus 4 makes 10'.

Repeat with different start numbers:

1.	7	4.	9	7.	4	9.	10
2.	2	5.	3	8.	1	10.	6
3.	5	6.	8				

Starter activity 78

Answers

Resources
A set of numeral cards 0–10 (enlarged from photocopiable page 70) for each child and for the teacher.

Objective
Know by heart all pairs of numbers with a total of 10 (eg 3 + 7).

1. 3
2. 8
3. 5
4. 1
5. 7
6. 2
7. 6
8. 9
9. 0
10. 4

Counting in ones, twos and tens

1. Count in ones from 0 to 30, then back again.

2. Count in twos from 0 to 30.

3. Count in twos from 1 to 29.

4. Count in tens from 0 to 100.

5. Count backwards in tens from 100.

6. Count on 3 tens from 40.

7. Count on 4 tens from 20.

8. Count in tens from 7 to 97.

9. Count in tens from 15 to 90.

10. Count backwards in tens from 93 to 3.

Starter activity 79

Resources
A 0–99 square (from photocopiable page 79) for each child.

Objective
Count on and back in ones from any small number, and in tens from and back to zero; count on in twos from zero, then 1.

Strategies
● Tell the children that they need to listen carefully. They are going to count in different ways, using the 0–99 square to help them.

Counting

Starter activity 80

Resources
A 0–20 number snake (from photocopiable page 75) for each child.

Objective
Describe and extend number sequences: count in twos from zero, then 1, and begin to recognise odd or even numbers to about 20 as 'every other number'.

1. even
2. odd
3. even
4. odd
5. even
6. odd

Using the snake, count together in twos from 0 to 20 and back again.

Ask: *What are the numbers called that we have said?* (Even numbers.) *Do you know where to start when you are counting the odd numbers?*

Using the snake, count together from 1 to 19 and back again.

Ask: *Are these odd or even numbers?*

1. 2, 4, 6

2. 1, 3, 5

3. 8, 10, 12

4. 7, 9, 11

5. 14, 16, 18, 20

6. 13, 15, 17, 19

Counting

Count together in twos from 0 to 20 and back again. Ask: *Have we counted the even numbers or the odd numbers?*

Count together in twos from 1 to 19 and back again. Ask: *Have we counted the odd numbers or the even numbers?*

Ask the children how they remember which numbers are odd and even. Collect several answers.

Count together in fives:
- from zero to 20 and back again
- from 5 to 20, then back to zero
- from 10 to 20, then back to zero.

Counting

Tell the children that they are going to count in threes. To do this, they have to 'step' two numbers each time.

Say: *Put a finger on zero.* Count together: 'Step, step, 3, step, step, 6...' Continue to 30. Repeat.

Count backwards together from 30: 'Step, step, 27, step, step, 24...'.

Count forwards together in threes, only saying the numbers. The children should touch the two 'step over' numbers each time.

Resources
A board or flip chart.

Objective
Describe and extend number sequences: **count on and back in ones from any small number, and in tens from and back to zero.**

Count in ones and tens

Draw a blank number line:

0   10

Ask volunteers to write the following numbers under the line:

1.	6	5.	7
2.	2	6.	1
3.	9	7.	5
4.	3	8.	8

Ask: *Which number is missing?* (4)

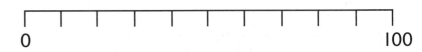

0   100

Draw a blank number line with the ends marked 0 and 100. Ask volunteers to write:

9.	40	13.	70
10.	80	14.	50
11.	10	15.	20
12.	30	16.	60

Ask: *Which number is missing?* (90)

Subtraction to 10

1. 6 – 2

2. 4 – 1

3. 10 – 7

4. 5 – 3

5. 9 – 6

6. 7 – 3

7. 5 – 1

8. 8 – 7

9. 6 – 6

10. 3 – 2

11. 10 – 5

12. 9 – 3

Starter activity 84

Answers

1. 4
2. 3
3. 3
4. 2
5. 3
6. 4
7. 4
8. 1
9. 0
10. 1
11. 5
12. 6

Resources
A number track 1–10 (photocopiable page 72) for each child.

Objective
Begin to know addition facts for all pairs of numbers with a total up to at least 10, and the corresponding subtraction facts.

Strategies
● Explain that you will ask the children to find the answer to a subtraction ('take away') question.
● Remind the children that they should 'hop' back from the start number, saying aloud the number of 'hops' that they make.

Starter activity 85

Resources
A board or flip chart.

Objective
Describe and extend number sequences: count in twos from zero, then one, and begin to recognise odd or even numbers to about 20 as 'every other number'.

Odd numbers

Draw a blank number line with nine equal divisions:

Explain to the children that they are going to write **odd** numbers under the line.

Ask: *Which number shall we start with?* Let a child write '1' under the first mark on the line.

Ask volunteers to write:

1.	5	5.	15
2.	11	6.	19
3.	3	7.	13
4.	9	8.	17

Ask: *Which number is missing?* (7)

Count together in twos from 1 to 19 and back again. Repeat.

Count in tens

Draw a number line:

0   100

Count together in tens from zero to 100 and back, pointing to each line in turn.

Ask individuals to point to the correct line when you say a number.

1.	20	4.	60	7.	40
2.	50	5.	10	8.	80
3.	90	6.	70	9.	30

Repeat until each child has had a turn.

> **Starter activity 86**
>
> **Resources**
> A board or flip chart.
>
> **Objective**
> **Count in tens from and back to zero.**

Addition to 10

Answers

1. 5
2. 9
3. 8
4. 7
5. 7
6. 4
7. 6
8. 8
9. 3
10. 6

Starter activity 87

Resources
A set of numeral cards 0–10 (enlarged from photocopiable page 70) for each child.

Objective
Begin to know addition facts for all pairs of numbers with a total up to at least 10.

Strategies
● Ask the children to find the answer to each addition question. Encourage them to hold the first number in their head and then count on. They should answer by holding up the appropriate numeral card.

1. $4 + 1$
2. $8 + 1$
3. $5 + 3$
4. $5 + 2$
5. $3 + 4$

6. $1 + 3$
7. $6 + 0$
8. $4 + 4$
9. $2 + 1$
10. $4 + 2$

Addition to 10

Answers

1. 3
2. 6
3. 6
4. 9
5. 4
6. 8
7. 7
8. 10
9. 9
10. 5

Starter activity 88

Resources
A set of numeral cards 0–10 (enlarged from photocopiable page 70) for each child.

Objective
Begin to know addition facts for all pairs of numbers with a total up to at least 10.

Strategies
● Ask the children to answer by holding up the appropriate numeral card.
● Encourage counting on as a strategy.

1. $1 + 2$
2. $3 + 3$
3. $5 + 1$
4. $7 + 2$
5. $2 + 2$

6. $8 + 0$
7. $4 + 3$
8. $9 + 1$
9. $5 + 4$
10. $3 + 2$

SCHOLASTIC

Subtraction to 10

1. 5 – 1

2. 5 – 3

3. 3 – 2

4. 4 – 1

5. 3 – 0

6. 4 – 2

7. 6 – 1

8. 6 – 3

9. 5 – 4

10. 4 – 3

Starter activity 89

Resources
A set of numeral cards 0–10 (enlarged from photocopiable page 70) for each child.

Objective
Begin to know addition facts for all pairs of numbers with a total up to at least 10, and the corresponding subtraction facts.

Strategies
• Explain that the children need to 'take away' to find the answer. Using their fingers may be helpful to them. They should answer by holding up the appropriate numeral card.

Answers

1. 4
2. 2
3. 1
4. 3
5. 3
6. 2
7. 5
8. 3
9. 1
10. 1

Subtraction to 10

1. 5 – 2

2. 4 – 4

3. 7 – 2

4. 7 – 4

5. 9 – 1

6. 9 – 5

7. 2 – 2

8. 4 – 0

9. 6 – 2

10. 5 – 5

Starter activity 90

Resources
A set of numeral cards 0–10 (enlarged from photocopiable page 70) for each child.

Objective
Begin to know addition facts for all pairs of numbers up to at least 10, and the corresponding subtraction facts.

Strategies
• Explain that the children need to 'take away' to find the answer. They may use their fingers to help them. Ask them to answer by holding up the appropriate numeral card.

Answers

1. 3
2. 0
3. 5
4. 3
5. 8
6. 4
7. 0
8. 4
9. 4
10. 0

Starter activity 91

Resources
A 0–99 square (from photocopiable page 79) for each child.

Objective
Describe and extend number sequences: count in steps of 5 from zero to 20 or more, then back again.

Counting in fives

Explain to the children that they are going to count in fives. Ask the class to count together in fives to 30, using the 0–99 square. Repeat.

Discuss the patterns that the children can see on the 0–99 square when they count in fives.

Count together in fives to 30. Repeat. Children who feel confident can turn over the 0–99 square and count in their heads.

Count around the class in fives to 30, then count back to zero together. Repeat.

Starter activity 92

Resources
A 0–99 square (from photocopiable page 79) for each child.

Objective
Describe and extend number sequences: count in steps of 5 from zero to 20 or more, then back again.

Strategies
• Using the 0–99 square, count together in fives from zero to 30 and back again. Repeat.
• Count round the class in fives from zero to 30 and back again.
• Ask: *Who can say the next number in the pattern?*

Answers

1. 20
2. 25
3. 30
4. 15
5. 10
6. 5

Counting in fives

1.   5   10   15   ?

2.   10   15   20   ?

3.   15   20   25   ?

4.   30   25   20   ?

5.   25   20   15   ?

6.   20   15   10   ?

■SCHOLASTIC

Counting in threes

Explain that the children are going to count in threes by stepping over two numbers each time.

Count together up to 30, saying: 'Step, step, 3, step, step, 6...'. Repeat.

Count in threes to 30 together, saying only the numbers, with the children touching the two 'steps' each time. Repeat.

Starter activity 93

Resources
A 0–99 square (from photocopiable page 79) for each child.

Objective
Describe and extend number sequences: begin to count in steps of 3 from zero.

Doubling

Hold up a 'Domino double' card and say what it is (eg double 2). The children should raise a hand to say the number that is equal to it (4). Ask an individual to stand beside you with the appropriate numeral card. All say: 'Double 2 is 4.'

What number is this equal to?

1.   double 3
2.   double 1
3.   double 5

4.   double 2
5.   double 0
6.   double 4

Starter activity 94

Resources
A set of 'Domino doubles' cards (enlarged from photocopiable page 74); a set of even-numbered numeral cards 0–10 (enlarged from photocopiable page 70).

Objective
Know by heart addition doubles of all numbers to at least 5 (eg 4 + 4).

Strategies
● Repeat the whole activity if time allows.

Answers
1. 6
2. 2
3. 10
4. 4
5. 0
6. 8

Doubling

Starter activity 95

Resources
A set of 'Domino doubles' cards (enlarged from photocopiable page 74); a set of even-numbered numeral cards 0–10 (enlarged from photocopiable page 70).

Objective
Know by heart addition doubles of all numbers to at least 5 (eg 4 + 4).

Spread out the Domino doubles face up on the carpet.

Hold up a numeral card (eg 4) and ask which Domino double is equal to it (double 2). The children should raise a hand to answer.

Ask an individual to stand beside you with the appropriate domino card. All say: 'Double 2 equals 4.'

Repeat the activity until all the numeral cards have been used at least twice.

Answers

1. 4
2. 7
3. 7
4. 3
5. 6
6. 8
7. 9
8. 4
9. 8
10. 9

Starter activity 96

Resources
A set of numeral cards 0–10 (enlarged from photocopiable page 70) for each child.

Objective
Begin to know addition facts for all pairs of numbers with a total up to at least 10, and the corresponding subtraction facts.

Strategies
• The children answer by holding up a numeral card.

Addition to 10

1. 3 add 1

2. How many are 2 and 5 together?

3. 4 plus 3

4. What must I add to 7 to make 10?

5. What must I add to 4 to make 10?

6. 6 plus 2

7. 3 add 6

8. 2 add 2

9. double 4

10. Add 5 to 4.

Addition to 10

1. 4 add 1

2. 3 plus 5

3. What must I add to 5 to make 10?

4. What must I add to 2 to make 10?

5. Add 4 to 2.

6. 7 plus 0

7. 2 add 7

8. How many are 1 and 8 together?

9. double 3

10. 0 plus 9

Starter activity 97

Resources
A set of numeral cards 0–10 (enlarged from photocopiable page 70) for each child.

Objective
Begin to know addition facts for all pairs of numbers with a total up to at least 10, and the corresponding subtraction facts.

Strategies
• Ask the children to answer by holding up a numeral card.

Answers
1. 5
2. 8
3. 5
4. 8
5. 6
6. 7
7. 9
8. 9
9. 6
10. 9

Telling the time

1. 4 o'clock

2. half past 4

3. 12 o'clock

4. 9 o'clock

5. half past 2

6. half past 5

7. half past 10

8. 6 o'clock

9. How long is it from 6 o'clock to 7 o'clock?

10. How long is it from 3 o'clock to 4 o'clock?

11. How long is it from 6 o'clock to 8 o'clock?

12. How long is it from half past 2 to half past 3?

Starter activity 99

Resources
A teaching clock.

Objective
Read the time to the hour or half-hour on analogue clocks.

Strategies
• Ask the children to tell you the times as you show them on the clock face.
• Ask questions 9–12 without demonstrating them on the clock.

Starter activity 98

Resources
A teaching clock.

Objective
Read the time to the hour or half-hour on analogue clocks.

Telling the time

Ask the children to tell you times as you show them on the clock face.

1.	2 o'clock	4.	8 o'clock
2.	5 o'clock	5.	10 o'clock
3.	1 o'clock	6.	6 o'clock

Ask: *What time is it?* as you show the children 9 o'clock, then half past 9. If necessary, explain that half past 9 is half an hour after 9 o'clock.

What time is it?

7.	3 o'clock	10.	half past 7
8.	half past 3	11.	11 o'clock
9.	7 o'clock	12.	half past 11

Picnic 2 pm

Bridging 10 and 20

Starter activity 100

Resources
A board or flip chart.

Objective
Begin to bridge through 10, and later 20, when adding a single-digit number.

Write:

$$18 + 5 =$$

Ask: *How can we work it out?* Encourage the children to suggest methods, such as: '18 add 2 is 20, and add 3 more is 23.'

Write:

$$18 + 5 = 18 + 2 + 3 = 20 + 3 = 23$$

Remind the children that they are trying to make 20 because adding on to 20 is easy!

Write each of the examples in turn, and ask an individual to say the answer and explain how he or she worked it out. Write out each partitioned equation in full, as above.

1. $17 + 8$
2. $15 + 7$
3. $16 + 6$
4. $19 + 4$
5. $18 + 6$

Numeral cards 0–10

0	1	2
3	4	5
6	7	8
9	10	=

Numeral cards 11–20

11	12	13
14	15	16
17	18	19
20	+	−

Number tracks 1–10

1	2	3	4	5	6	7	8	9	10
1	2	3	4	5	6	7	8	9	10
1	2	3	4	5	6	7	8	9	10
1	2	3	4	5	6	7	8	9	10

Tens number lines 0–100

0 10 20 30 40 50 60 70 80 90 100

0 10 20 30 40 50 60 70 80 90 100

0 10 20 30 40 50 60 70 80 90 100

0 10 20 30 40 50 60 70 80 90 100

Domino doubles

Number snake 0–20

Make 10 cards

Enlarge this page to A3 and cut it in half to make two A4 sheets. Enlarge each A4 sheet to A3 on card. Cut out each fact card.

$\square + 0 = 10$	$0 + \square = 10$
$\square + 1 = 10$	$1 + \square = 10$
$\square + 2 = 10$	$2 + \square = 10$
$\square + 3 = 10$	$3 + \square = 10$
$\square + 4 = 10$	$4 + \square = 10$
$\square + 5 = 10$	$5 + \square = 10$
$\square + 6 = 10$	$6 + \square = 10$
$\square + 7 = 10$	$7 + \square = 10$
$\square + 8 = 10$	$8 + \square = 10$
$\square + 9 = 10$	$9 + \square = 10$
$\square + 10 = 10$	$10 + \square = 10$

Fact cards: addition

✂

0 + 1	2 + 0
0 + 2	2 + 1
0 + 3	2 + 2
0 + 4	2 + 3
0 + 5	3 + 0
1 + 0	3 + 1
1 + 1	3 + 2
1 + 2	4 + 0
1 + 3	4 + 1
1 + 4	5 + 0

Fact cards: subtraction

5 – 0	4 – 4
5 – 1	3 – 0
5 – 2	3 – 1
5 – 3	3 – 2
5 – 4	3 – 3
5 – 5	2 – 0
4 – 0	2 – 1
4 – 1	2 – 2
4 – 2	1 – 0
4 – 3	1 – 1

0–99 square

0	1	2	3	4	5	6	7	8	9
10	11	12	13	14	15	16	17	18	19
20	21	22	23	24	25	26	27	28	29
30	31	32	33	34	35	36	37	38	39
40	41	42	43	44	45	46	47	48	49
50	51	52	53	54	55	56	57	58	59
60	61	62	63	64	65	66	67	68	69
70	71	72	73	74	75	76	77	78	79
80	81	82	83	84	85	86	87	88	89
90	91	92	93	94	95	96	97	98	99

100 Mental Maths Starters

Year 1

Index

Note that the numbers given here are activity numbers, not page numbers.

Addition
adding 1   *11, 37*
adding 2   *11*
adding 10   *37, 51, 61, 62*
of 3 numbers   *52, 53*
pairs totalling 10   *21, 26, 29, 33, 39, 41, 45, 73, 75, 78*
to 5   *1, 4, 12, 27, 42, 49*
to 10   *8, 9, 11, 54, 55, 56, 63, 65, 68, 69, 87, 88, 96, 97*
to 20   *23, 24, 57, 58*

Bridging
10   *72, 74, 100*
20   *100*

Counting
counting on   *8, 9, 19, 20, 23, 24, 79*
in fives   *81, 91, 92*
in tens   *25, 32, 38, 79*
in threes   *82, 93*
in twos   *79, 80, 81*
objects   *1, 2, 10, 31, 36, 48*
to 10   *5, 7*
to 20   *13, 14, 17, 18, 47*
to 30   *37*

Doubles
of numbers to 5   *16, 30, 34, 35, 76, 77, 94, 95*
of numbers to 10   *44*

Even numbers   *46, 70, 71, 80, 81*

Near doubles   *77*

Odd numbers   *35, 46, 70, 71, 80, 81, 85*

Ordering
tens to 100   *38, 83, 86*
to 5   *6*
to 10   *3, 5, 7, 22, 83, 86*
to 20   *15, 70, 71, 85*

Partitioning
'5 and a bit'   *28, 40*
teens numbers   *59, 60, 66, 67*

Subtraction
taking away 10   *51, 61, 62*
to 5   *2, 27, 43, 50*
to 10   *54, 55, 56, 64, 65, 89, 90*

Time
reading o'clock and half past   *98, 99*